Name: ____________________ Date: ______

AF813106

Addition Facts

Say It! Write It! Repeat It!

1 + 0 = 1
1 + 1 = 2
1 + 2 = 3
1 + 3 = 4
1 + 4 = 5
1 + 5 = 6
1 + 6 = 7
1 + 7 = 8
1 + 8 = 9
1 + 9 = 10
1 + 10 = 11
1 + 11 = 12
1 + 12 = 13

2 + 0 = 2
2 + 1 = 3
2 + 2 = 4
2 + 3 = 5
2 + 4 = 6
2 + 5 = 7
2 + 6 = 8
2 + 7 = 9
2 + 8 = 10
2 + 9 = 11
2 + 10 = 12
2 + 11 = 13
2 + 12 = 14

3 + 0 = 3
3 + 1 = 4
3 + 2 = 5
3 + 3 = 6
3 + 4 = 7
3 + 5 = 8
3 + 6 = 9
3 + 7 = 10
3 + 8 = 11
3 + 9 = 12
3 +10 = 13
3 + 11 = 14
3 + 12 = 15

4 + 0 = 4
4 + 1 = 5
4 + 2 = 6
4 + 3 = 7
4 + 4 = 8
4 + 5 = 9
4 + 6 = 10
4 + 7 = 11
4 + 8 = 12
4 + 9 = 13
4 + 10 = 14
4 + 11 = 15
4 + 12 = 16

5 + 0 = 5
5 + 1 = 6
5 + 2 = 7
5 + 3 = 8
5 + 4 = 9
5 + 5 = 10
5 + 6 = 11
5 + 7 = 12
5 + 8 = 13
5 + 9 = 14
5 + 10 = 15
5 + 11 = 16
5 + 12 = 17

6 + 0 = 6
6 + 1 = 7
6 + 2 = 8
6 + 3 = 9
6 + 4 = 10
6 + 5 = 11
6 + 6 = 12
6 + 7 = 13
6 + 8 = 14
6 + 9 = 15
6 + 10 = 16
6 + 11 = 17
6 + 12 = 18

7 + 0 = 7
7 + 1 = 8
7 + 2 = 9
7 + 3 = 10
7 + 4 = 11
7 + 5 = 12
7 + 6 = 13
7 + 7 = 14
7 + 8 = 15
7 + 9 = 16
7 + 10 = 17
7 + 11 = 18
7 + 12 = 19

8 + 0 = 8
8 + 1 = 9
8 + 2 = 10
8 + 3 = 11
8 + 4 = 12
8 + 5 = 13
8 + 6 = 14
8 + 7 = 15
8 + 8 = 16
8 + 9 = 17
8 + 10 = 18
8 + 11 = 19
8 + 12 = 20

9 + 0 = 9
9 + 1 = 10
9 + 2 = 11
9 + 3 = 12
9 + 4 = 13
9 + 5 = 14
9 + 6 = 15
9 + 7 = 16
9 + 8 = 17
9 + 9 = 18
9 + 10 = 19
9 + 11 = 20
9 + 12 = 21

10 + 0 = 10
10 + 1 = 11
10 + 2 = 12
10 + 3 = 13
10 + 4 = 14
10 + 5 = 15
10 + 6 = 16
10 + 7 = 17
10 + 8 = 18
10 + 9 = 19
10 + 10 = 20
10 + 11 = 21
10 + 12 = 22

11 + 0 = 11
11 + 1 = 12
11 + 2 = 13
11 + 3 = 14
11 + 4 = 15
11 + 5 = 16
11 + 6 = 17
11 + 7 = 18
11 + 8 = 19
11 + 9 = 20
11 + 10 = 21
11 + 11 = 22
11 + 12 = 23

12 + 0 = 12
12 + 1 = 13
12 + 2 = 14
12 + 3 = 15
12 + 4 = 16
12 + 5 = 17
12 + 6 = 18
12 + 7 = 19
12 + 8 = 20
12 + 9 = 21
12 + 10 = 22
12 + 11 = 23
12 + 12 = 24

Time: __________

Name: ______________________ Date: ______________

Addition Facts

Say It!

1 + 1 = 2	1 + 4 = 5	1 + 7 = 8	1 + 10 = 11
1 + 2 = 3	1 + 5 = 6	1 + 8 = 9	1 + 11 = 12
1 + 3 = 4	1 + 6 = 7	1 + 9 = 10	1 + 12 = 13

Write It! Repeat It!

1 + 1 = ___	1 + 10= ___	1 + 9 = ___	1 + 8 = ___
1 + 4 = ___	1 + 2 = ___	1 + 6 = ___	1 + 12= ___
1 + 6 = ___	1 + 11= ___	1 + 3 = ___	1 + 2 = ___
1 + 9 = ___	1 + 11= ___	1 + 1 = ___	1 + 11= ___
1 + 11= ___	1 + 5 = ___	1 + 4 = ___	1 + 9 = ___
1 + 5 = ___	1 + 3 = ___	1 + 5 = ___	1 + 6 = ___
1 + 3 = ___	1 + 1 = ___	1 + 3 = ___	1 + 3 = ___
1 + 7 = ___	1 + 4 = ___	1 + 4 = ___	1 + 1 = ___
1 + 8 = ___	1 + 5 = ___	1 + 6 = ___	1 + 4 = ___
1 + 12= ___	1 + 3 = ___	1 + 9 = ___	1 + 7 = ___
1 + 2 = ___	1 + 1 = ___	1 + 11= ___	1 + 8 = ___

Time: ___________

Name: Date:

Addition Facts

Say It!

2 + 1 = 3 2 + 2 = 4 2 + 3 = 5	2 + 4 = 6 2 + 5 = 7 2 + 6 = 8	2 + 7 = 9 2 + 8 = 10 2 + 9 = 11	2 + 10 = 12 2 + 11 = 13 2 + 12 = 14

Write It! Repeat It!

2 + 3 = ___	2 + 1 = ___	2 + 9 = ___	2 + 8 = ___
2 + 7 = ___	2 + 5 = ___	2 + 10= ___	2 + 9 = ___
2 + 6 = ___	2 + 9 = ___	2 + 12= ___	2 + 1 = ___
2 + 2 = ___	2 + 10= ___	2 + 8 = ___	2 + 5 = ___
2 + 11= ___	2 + 12= ___	2 + 9 = ___	2 + 9 = ___
2 + 4= ___	2 + 8 = ___	2 + 7 = ___	2 + 10= ___
2 + 5 = ___	2 + 9 = ___	2 + 6 = ___	2 + 12= ___
2 + 8 = ___	2 + 7 = ___	2 + 2 = ___	2 + 8 = ___
2 + 9 = ___	2 + 6 = ___	2 + 11= ___	2 + 9 = ___
2 + 10= ___	2 + 2 = ___	2 + 10= ___	2 + 10= ___
2 + 12= ___	2 + 11= ___	2 + 12= ___	2 + 12= ___

Time:

Name: ______________________ Date: ______________

Addition Facts

Say It!

3 + 1 = 4 3 + 2 = 5 3 + 3 = 6	3 + 4 = 7 3 + 5 = 8 3 + 6 = 9	3 + 7 = 10 3 + 8 = 11 3 + 9 = 12	3 +10 = 13 3 + 11 = 14 3 + 12 = 15

Write It! Repeat It!

3 + 5 = ___	3 + 12= ___	3 + 7 = ___	3 + 11 = ___
3 + 2 = ___	3 +10 = ___	3 + 1 = ___	3 + 6 = ___
3 + 3 = ___	3 + 11 = ___	3 + 8 = ___	3 + 5 = ___
3 + 4 = ___	3 + 6 = ___	3 + 2 = ___	3 + 2 = ___
3 + 9 = ___	3 + 5 = ___	3 + 3 = ___	3 + 3 = ___
3 +10 = ___	3 + 2 = ___	3 + 4 = ___	3 + 4 = ___
3 + 11 = ___	3 + 3 = ___	3 + 9 = ___	3 + 12= ___
3 + 6 = ___	3 + 4 = ___	3 +10 = ___	3 +10 = ___
3 + 7 = ___	3 + 11 = ___	3 + 11 = ___	3 + 7 = ___
3 + 1 = ___	3 + 6 = ___	3 + 6 = ___	3 + 1 = ___
3 + 8 = ___	3 + 7 = ___	3 + 2 = ___	3 + 8 = ___

Time: ___________

Name: ______________________ Date: ______________

Addition Facts

Say It!

4 + 1 = 5	4 + 4 = 8	4 + 7 = 11	4 + 10 = 14
4 + 2 = 6	4 + 5 = 9	4 + 8 = 12	4 + 11 = 15
4 + 3 = 7	4 + 6 = 10	4 + 9 = 13	4 + 12 = 16

Write It! Repeat It!

4 + 1 = ___	4 + 5 = ___	4 + 12= ___	4 + 5 = ___
4 + 9 = ___	4 + 2 = ___	4 + 10= ___	4 + 2 = ___
4 + 12= ___	4 + 3 = ___	4 + 9 = ___	4 + 3 = ___
4 + 10= ___	4 + 8 = ___	4 + 11 = ___	4 + 8 = ___
4 + 4 = ___	4 + 6 = ___	4 + 9 = ___	4 + 6 = ___
4 + 5 = ___	4 + 7 = ___	4 + 4 = ___	4 + 7 = ___
4 + 2 = ___	4 + 11 = ___	4 + 2 = ___	4 + 8 = ___
4 + 3 = ___	4 + 9 = ___	4 + 3 = ___	4 + 9 = ___
4 + 8 = ___	4 + 4 = ___	4 + 8 = ___	4 + 12= ___
4 + 6 = ___	4 + 5 = ___	4 + 4 = ___	4 + 10= ___
4 + 7 = ___	4 + 2 = ___	4 + 5 = ___	4 + 4 = ___

Time: ___________

Name: ______________________ Date: ____________

Addition Facts

Say It!

5 + 1 = 6 5 + 2 = 7 5 + 3 = 8	5 + 4 = 9 5 + 5 = 10 5 + 6 = 11	5 + 7 = 12 5 + 8 = 13 5 + 9 = 14	5 + 10 = 15 5 + 11 = 16 5 + 12 = 17

Write It! Repeat It!

5 + 8 = ___	5 + 12= ___	5 + 6 = ___	5 + 11= ___
5 + 2 = ___	5 + 7 = ___	5 + 2 = ___	5 + 7 = ___
5 + 3 = ___	5 + 6 = ___	5 + 3 = ___	5 + 11= ___
5 + 7 = ___	5 + 10= ___	5 + 7 = ___	5 + 9 = ___
5 + 9 = ___	5 + 2 = ___	5 + 4 = ___	5 + 6 = ___
5 + 4 = ___	5 + 3 = ___	5 + 8 = ___	5 + 10= ___
5 + 1 = ___	5 + 7 = ___	5 + 9 = ___	5 + 2 = ___
5 + 5 = ___	5 + 11= ___	5 + 4 = ___	5 + 3 = ___
5 + 6 = ___	5 + 9 = ___	5 + 1 = ___	5 + 7 = ___
5 + 10= ___	5 + 1 = ___	5 + 3 = ___	5 + 9 = ___
5 + 11= ___	5 + 5 = ___	5 + 7 = ___	5 + 12= ___

Time: ____________

Name: ______________________ Date: ______________

Addition Facts

Say It!

6 + 1 = 7	6 + 4 = 10	6 + 7 = 13	6 + 10 = 16
6 + 2 = 8	6 + 5 = 11	6 + 8 = 14	6 + 11 = 17
6 + 3 = 9	6 + 6 = 12	6 + 9 = 15	6 + 12 = 18

Write It! Repeat It!

6 + 12=___	6 + 8 =___	6 + 4 =___	6 + 2 =___
6 + 1 =___	6 + 3 =___	6 + 9 =___	6 + 11 =___
6 + 5 =___	6 + 2 =___	6 + 10=___	6 + 4 =___
6 + 3 =___	6 + 11 =___	6 + 12=___	6 + 9 =___
6 + 9 =___	6 + 4 =___	6 + 1 =___	6 + 8 =___
6 + 10=___	6 + 9 =___	6 + 5 =___	6 + 3 =___
6 + 6 =___	6 + 10=___	6 + 3 =___	6 + 2 =___
6 + 7 =___	6 + 12=___	6 + 9 =___	6 + 8 =___
6 + 2 =___	6 + 1 =___	6 + 8 =___	6 + 3 =___
6 + 11 =___	6 + 5 =___	6 + 3 =___	6 + 2 =___
6 + 4 =___	6 + 3 =___	6 + 2 =___	6 + 11 =___

Time: ____________

Name: ______________________ Date: ______________

Addition Facts

Say It!

7 + 1 = 8	7 + 4 = 11	7 + 7 = 14	7 + 10 = 17
7 + 2 = 9	7 + 5 = 12	7 + 8 = 15	7 + 11 = 18
7 + 3 = 10	7 + 6 = 13	7 + 9 = 16	7 + 12 = 19

Write It! Repeat It!

7 + 9 = ___	7 + 6 = ___	7 + 8 = ___	7 + 3 = ___
7 + 10= ___	7 + 12= ___	7 + 6 = ___	7 + 1 = ___
7 + 5 = ___	7 + 4 = ___	7 + 4 = ___	7 + 7 = ___
7 + 3 = ___	7 + 1 = ___	7 + 1 = ___	7 + 6 = ___
7 + 1 = ___	7 + 7 = ___	7 + 7 = ___	7 + 12= ___
7 + 7 = ___	7 + 12= ___	7 + 12= ___	7 + 4 = ___
7 + 11= ___	7 + 2 = ___	7 + 9 = ___	7 + 6 = ___
7 + 8 = ___	7 + 11= ___	7 + 10= ___	7 + 4 = ___
7 + 6 = ___	7 + 9 = ___	7 + 5 = ___	7 + 1 = ___
7 + 12= ___	7 + 10= ___	7 + 8 = ___	7 + 11= ___
7 + 2 = ___	7 + 5 = ___	7 + 6 = ___	7 + 9 = ___

Time: ____________

Name: ____________________ Date: ____________

Addition Facts

Say It!

8 + 1 = 9	8 + 4 = 12	8 + 7 = 15	8 + 10 = 18
8 + 2 = 10	8 + 5 = 13	8 + 8 = 16	8 + 11 = 19
8 + 3 = 11	8 + 6 = 14	8 + 9 = 17	8 + 12 = 20

Write It! Repeat It!

8 + 5 = ___	8 + 10= ___	8 + 4 = ___	8 + 3 = ___
8 + 8 = ___	8 + 1 = ___	8 + 2 = ___	8 + 8 = ___
8 + 4 = ___	8 + 7 = ___	8 + 5 = ___	8 + 1 = ___
8 + 2 = ___	8 + 12= ___	8 + 8 = ___	8 + 7 = ___
8 + 9 = ___	8 + 3 = ___	8 + 4 = ___	8 + 10= ___
8 + 11= ___	8 + 6 = ___	8 + 9 = ___	8 + 1 = ___
8 + 3 = ___	8 + 10= ___	8 + 7 = ___	8 + 12= ___
8 + 6 = ___	8 + 2 = ___	8 + 12= ___	8 + 4 = ___
8 + 10= ___	8 + 9 = ___	8 + 3 = ___	8 + 8 = ___
8 + 1 = ___	8 + 11= ___	8 + 6 = ___	8 + 4 = ___
8 + 7 = ___	8 + 8 = ___	8 + 11= ___	8 + 9 = ___

Time: __________

Name: ____________________ Date: ____________

Addition Facts

Say It!

9 + 1 = 10	9 + 4 = 13	9 + 7 = 16	9 + 10 = 19
9 + 2 = 11	9 + 5 = 14	9 + 8 = 17	9 + 11 = 20
9 + 3 = 12	9 + 6 = 15	9 + 9 = 18	9 + 12 = 21

Write It! Repeat It!

9 + 1 =___	9 + 5 =___	9 + 11 =___	9 + 4 =___
9 + 3 =___	9 + 1 =___	9 + 12=___	9 + 2 =___
9 + 11 =___	9 + 4 =___	9 + 4 =___	9 + 1 =___
9 + 6 =___	9 + 2 =___	9 + 2 =___	9 + 4 =___
9 + 7 =___	9 + 12=___	9 + 6 =___	9 + 2 =___
9 + 8 =___	9 + 4 =___	9 + 7 =___	9 + 12=___
9 + 12=___	9 + 2 =___	9 + 5 =___	9 + 4 =___
9 + 4 =___	9 + 9 =___	9 + 4 =___	9 + 2 =___
9 + 2 =___	9 + 3 =___	9 + 2 =___	9 + 3 =___
9 + 9 =___	9 + 11 =___	9 + 9 =___	9 + 11 =___
9 + 10=___	9 + 6 =___	9 + 1 =___	9 + 6 =___

Time:____________

Name: ______________________ Date: ______________

Addition Facts

Say It!

10 + 1 = 11 10 + 2 = 12 10 + 3 = 13	10 + 4 = 14 10 + 5 = 15 10 + 6 = 16	10 + 7 = 17 10 + 8 = 18 10 + 9 = 19	10 + 10 = 20 10 + 11 = 21 10 + 12 = 22

Write It! Repeat It!

10 + 1 = ___	10 + 10= ___	10 + 8 = ___	10 + 10= ___
10 + 5 = ___	10 + 1 = ___	10 + 6 = ___	10 + 1 = ___
10 + 2 = ___	10 + 8 = ___	10 + 12= ___	10 + 8 = ___
10 + 4 = ___	10 + 6 = ___	10 + 1 = ___	10 + 1 = ___
10 + 8 = ___	10 + 12= ___	10 + 7 = ___	10 + 5 = ___
10 + 6 = ___	10 + 9 = ___	10 + 11= ___	10 + 2 = ___
10 + 12= ___	10 + 3 = ___	10 + 5 = ___	10 + 4 = ___
10 + 2 = ___	10 + 5 = ___	10 + 2 = ___	10 + 12= ___
10 + 3 = ___	10 + 2 = ___	10 + 10= ___	10 + 9 = ___
10 + 9 = ___	10 + 7 = ___	10 + 1 = ___	10 + 3 = ___
10 + 3 = ___	10 + 11= ___	10 + 8 = ___	10 + 8 = ___

Time: __________

Name: ______________________ Date: ______________

Addition Facts

Say It!

11 + 1 = 12 11 + 2 = 13 11 + 3 = 14	11 + 4 = 15 11 + 5 = 16 11 + 6 = 17	11 + 7 = 18 11 + 8 = 19 11 + 9 = 20	11 + 10 = 21 11 + 11 = 22 11 + 12 = 23

Write It! Repeat It!

11 + 1 = ____	11 + 5 = ____	11 + 4 = ____	11 + 10= ____
11 + 2 = ____	11 + 7 = ____	11 + 1 = ____	11 + 8 = ____
11 + 4 = ____	11 + 11= ____	11 + 2 = ____	11 + 6 = ____
11 + 8 = ____	11 + 3 = ____	11 + 8 = ____	11 + 7 = ____
11 + 6 = ____	11 + 9 = ____	11 + 6 = ____	11 + 11= ____
11 + 7 = ____	11 + 10= ____	11 + 7 = ____	11 + 3 = ____
11 + 11= ____	11 + 12= ____	11 + 6 = ____	11 + 9 = ____
11 + 3 = ____	11 + 1 = ____	11 + 12= ____	11 + 1 = ____
11 + 9 = ____	11 + 2 = ____	11 + 1 = ____	11 + 2 = ____
11 + 10= ____	11 + 4 = ____	11 + 2 = ____	11 + 4 = ____
11 + 12= ____	11 + 7 = ____	11 + 9 = ____	11 + 5 = ____

Time: __________

Name: ____________________ Date: ______________

Addition Facts

Say It!

12 + 1 = 13	12 + 4 = 16	12 + 7 = 19	12 + 10 = 22
12 + 2 = 14	12 + 5 = 17	12 + 8 = 20	12 + 11 = 23
12 + 3 = 15	12 + 6 = 18	12 + 9 = 21	12 + 12 = 24

Write It! Repeat It!

12 + 7 = ___	12 + 12= ___	12 + 10= ___	12 + 4 = ___
12 + 3 = ___	12 + 2 = ___	12 + 3 = ___	12 + 6 = ___
12 + 6 = ___	12 + 3 = ___	12 + 6 = ___	12 + 7 = ___
12 + 9 = ___	12 + 6 = ___	12 + 7 = ___	12 + 3 = ___
12 + 10= ___	12 + 7 = ___	12 + 6 = ___	12 + 2 = ___
12 + 4 = ___	12 + 3 = ___	12 + 1 = ___	12 + 5 = ___
12 + 1 = ___	12 + 8 = ___	12 + 11= ___	12 + 9 = ___
12 + 11= ___	12 + 6 = ___	12 + 2 = ___	12 + 6 = ___
12 + 12= ___	12 + 2 = ___	12 + 5 = ___	12 + 3 = ___
12 + 2 = ___	12 + 5 = ___	12 + 9 = ___	12 + 12= ___
12 + 5 = ___	12 + 9 = ___	12 + 10= ___	12 + 10= ___

Time: ____________

Name: ______________ Date: ______________

Addition to Subtraction

Say It! Write It! Repeat It!

1 + 1 = 2 - 1 = 1

1 + 2 = 3 - 2 = 1

1 + 3 = 4 - 3 = 1

1 + 4 = 5 - 4 = 1

1 + 5 = 6 - 5 = 1

1 + 6 = 7 - 6 = 1

1 + 7 = 8 - 7 = 1

1 + 8 = 9 - 8 = 1

1 + 9 = 10 - 9 = 1

1 + 10 = 11 - 10 = 1

1 + 11 = 12 - 11 = 1

1 + 12 = 13 - 12 = 1

Time: ______________

Name: ______________________ Date: ______________

Addition to Subtraction

Say It! Write It! Repeat It!

2 + 1 = 3 - 1 = 2

2 + 5 = 7 - 5 = 2

2 + 9 = 11 - 9 = 2

2 + 2 = 4 - 2 = 2

2 + 6 = 8 - 6 = 2

2 + 10 = 12 - 10 = 2

2 + 3 = 5 - 3 = 2

2 + 7 = 9 - 7 = 2

2 + 11 = 13 - 11 = 2

2 + 4 = 6 - 4 = 2

2 + 8 = 10 - 8 = 2

2 + 12 = 14 - 12 = 2

Time: __________

Name: ____________ Date: ____________

Addition to Subtraction

Say It! Write It! Repeat It!

3 + 1 = 4 - 1 = 3

3 + 5 = 8 - 5 = 3

3 + 9 = 12 - 9 = 3

3 + 2 = 5 - 2 = 3

3 + 6 = 9 - 6 = 3

3 +10 = 13 - 10 = 3

3 + 3 = 6 - 3 = 3

3 + 7 = 10 - 7 = 3

3 + 11 = 14 - 11 = 3

3 + 4 = 7 - 4 = 3

3 + 8 = 11 - 8 = 3

3 + 12 = 15 - 12 = 3

Time: ____________

Name: ______________________ Date: ______________

Addition to Subtraction

Say It! Write It! Repeat It!

4 + 1 = 5 - 1 = 4

4 + 5 = 9 - 5 = 4

4 + 9 = 13 - 9 = 4

4 + 2 = 6 - 2 = 4

4 + 6 = 10 - 6 = 4

4 + 10 = 14 - 10 = 4

4 + 3 = 7 - 3 = 4

4 + 7 = 11 - 7 = 4

4 + 11 = 15 - 11 = 4

4 + 4 = 8 - 4 = 4

4 + 8 = 12 - 8 = 4

4 + 12 = 16 - 12 = 4

Time: ____________

Name: ____________________ Date: ____________

Addition to Subtraction

Say It! Write It! Repeat It!

5 + 1 = 6 - 1 = 5

5 + 2 = 7 - 2 = 5

5 + 3 = 8 - 3 = 5

5 + 4 = 9 - 4 = 5

5 + 5 = 10 - 5 = 5

5 + 6 = 11 - 6 = 5

5 + 7 = 12 - 7 = 5

5 + 8 = 13 - 8 = 5

5 + 9 = 14 - 9 = 5

5 + 10 = 15 - 10 = 5

5 + 11 = 16 - 11 = 5

5 + 12 = 17 - 12 = 5

Time: __________

Name: ______________________ Date: ______________

Addition to Subtraction

Say It! Write It! Repeat It!

6 + 1 = 7 - 1 = 6

6 + 2 = 8 - 2 = 6

6 + 3 = 9 - 3 = 6

6 + 4 = 10 - 4 = 6

6 + 5 = 11 - 5 = 6

6 + 6 = 12 - 6 = 6

6 + 7 = 13 - 7 = 6

6 + 8 = 14 - 8 = 6

6 + 9 = 15 - 9 = 6

6 + 10 = 16 - 10 = 6

6 + 11 = 17 - 11 = 6

6 + 12 = 18 - 12 = 6

Time: __________

Name: ____________ Date: ____________

Addition to Subtraction

Say It! Write It! Repeat It!

7 + 1 = 8 - 1 = 7

7 + 5 = 12 - 5 = 7

7 + 9 = 16 - 9 = 7

7 + 2 = 9 - 2 = 7

7 + 6 = 13 - 6 = 7

7 + 10 = 17 - 10 = 7

7 + 3 = 10 - 3 = 7

7 + 7 = 14 - 7 = 7

7 + 11 = 18 - 11 = 7

7 + 4 = 11 - 4 = 7

7 + 8 = 15 - 8 = 7

7 + 12 = 19 - 12 = 7

Time: ____________

Name: ______________________ Date: ______________

Addition to Subtraction

Say It! Write It! Repeat It!

8 + 1 = 9 - 1 = 8

8 + 5 = 13 - 5 = 8

8 + 9 = 17 - 9 = 8

8 + 2 = 10 - 2 = 8

8 + 6 = 14 - 6 = 8

8 + 10 = 18 - 10 = 8

8 + 3 = 11 - 3 = 8

8 + 7 = 15 - 7 = 8

8 + 11 = 19 - 11 = 8

8 + 4 = 12 - 4 = 8

8 + 8 = 16 - 8 = 8

8 + 12 = 20 - 12 = 8

Time: ____________

Name: ____________________ Date: ____________

Addition to Subtraction

Say It! Write It! Repeat It!

9 + 1 = 10 - 1 = 9

9 + 2 = 11 - 2 = 9

9 + 3 = 12 - 3 = 9

9 + 4 = 13 - 4 = 9

9 + 5 = 14 - 5 = 9

9 + 6 = 15 - 6 = 9

9 + 7 = 16 - 7 = 9

9 + 8 = 17 - 8 = 9

9 + 9 = 18 - 9 = 9

9 + 10 = 19 - 10 = 9

9 + 11 = 20 - 11 = 9

9 + 12 = 21 - 12 = 9

Time: __________

Name: ____________ Date: ____________

Addition to Subtraction

Say It! Write It! Repeat It!

10 + 1 = 11 - 1 = 10

10 + 2 = 12 - 2 = 10

10 + 3 = 13 - 3 = 10

10 + 4 = 14 - 4 = 10

10 + 5 = 15 - 5 = 10

10 + 6 = 16 - 6 = 10

10 + 7 = 17 - 7 = 10

10 + 8 = 18 - 8 = 10

10 + 9 = 19 - 9 = 10

10 + 10 = 20 - 10 = 10

10 + 11 = 21 - 11 = 10

10 + 12 = 22 - 12 = 10

Time: ____________

Name: ______________________ Date: ______________

Addition to Subtraction

Say It! Write It! Repeat It!

11 + 1 = 12 - 1 = 11	11 + 5 = 16 - 5 = 11	11 + 9 = 20 - 9 = 11
11 + 2 = 13 - 2 = 11	11 + 6 = 17 - 6 = 11	11 + 10 = 21 - 10 = 11
11 + 3 = 14 - 3 = 11	11 + 7 = 18 - 7 = 11	11 + 11 = 22 - 11 = 11
11 + 4 = 15 - 4 = 11	11 + 8 = 19 - 8 = 11	11 + 12 = 23 - 12 = 11

Time: __________

Name: ______________________ Date: ______________

Addition to Subtraction

Say It! Write It! Repeat It!

12 + 1 = 13 - 1 = 12

12 + 5 = 17 - 5 = 12

12 + 9 = 21 - 9 = 12

12 + 2 = 14 - 2 = 12

12 + 6 = 18 - 6 = 12

12 + 10 = 22 - 10 = 12

12 + 3 = 15 - 3 = 12

12 + 7 = 19 - 7 = 12

12 + 11 = 23 - 11 = 12

12 + 4 = 16 - 4 = 12

12 + 8 = 20 - 8 = 12

12 + 12 = 24 - 12 = 12

Time: ____________

Name: ________________ Date: ________________

Addition to Subtraction:

Say It! Write It! Repeat It!

1 + 0 = 1 - 0 = 1
1 + 1 = 2 - 1 = 1
1 + 2 = 3 - 2 = 1
1 + 3 = 4 - 3 = 1
1 + 4 = 5 - 4 = 1
1 + 5 = 6 - 5 = 1
1 + 6 = 7 - 6 = 1
1 + 7 = 8 - 7 = 1
1 + 8 = 9 - 8 = 1
1 + 9 = 10 - 9 = 1
1 + 10 = 11 - 10 = 1
1 + 11 = 12 - 11 = 1
1 + 12 = 13 - 12 = 1

2 + 0 = 2 - 0 = 2
2 + 1 = 3 - 1 = 2
2 + 2 = 4 - 2 = 2
2 + 3 = 5 - 3 = 2
2 + 4 = 6 - 4 = 2
2 + 5 = 7 - 5 = 2
2 + 6 = 8 - 6 = 2
2 + 7 = 9 - 7 = 2
2 + 8 = 10 - 8 = 2
2 + 9 = 11 - 9 = 2
2 + 10 = 12 - 10 = 2
2 + 11 = 13 - 11 = 2
2 + 12 = 14 - 12 = 2

3 + 0 = 3 - 0 = 3
3 + 1 = 4 - 1 = 3
3 + 2 = 5 - 2 = 3
3 + 3 = 6 - 3 = 3
3 + 4 = 7 - 4 = 3
3 + 5 = 8 - 5 = 3
3 + 6 = 9 - 6 = 3
3 + 7 = 10 - 7 = 3
3 + 8 = 11 - 8 = 3
3 + 9 = 12 - 9 = 3
3 +10 = 13 - 10 = 3
3 + 11 = 14 - 11 = 3
3 + 12 = 15 - 12 = 3

4 + 0 = 4 - 0 = 4
4 + 1 = 5 - 1 = 4
4 + 2 = 6 - 2 = 4
4 + 3 = 7 - 3 = 4
4 + 4 = 8 - 4 = 4
4 + 5 = 9 - 5 = 4
4 + 6 = 10 - 6 = 4
4 + 7 = 11 - 7 = 4
4 + 8 = 12 - 8 = 4
4 + 9 = 13 - 9 = 4
4 + 10 = 14 - 10 = 4
4 + 11 = 15 - 11 = 4
4 + 12 = 16 - 12 = 4

5 + 0 = 5 - 0 = 5
5 + 1 = 6 - 1 = 5
5 + 2 = 7 - 2 = 5
5 + 3 = 8 - 3 = 5
5 + 4 = 9 - 4 = 5
5 + 5 = 10 - 5 = 5
5 + 6 = 11 - 6 = 5
5 + 7 = 12 - 7 = 5
5 + 8 = 13 - 8 = 5
5 + 9 = 14 - 9 = 5
5 + 10 = 15 - 10 = 5
5 + 11 = 16 - 11 = 5
5 + 12 = 17 - 12 = 5

6 + 0 = 6 - 0 = 6
6 + 1 = 7 - 1 = 6
6 + 2 = 8 - 2 = 6
6 + 3 = 9 - 3 = 6
6 + 4 = 10 - 4 = 6
6 + 5 = 11 - 5 = 6
6 + 6 = 12 - 6 = 6
6 + 7 = 13 - 7 = 6
6 + 8 = 14 - 8 = 6
6 + 9 = 15 - 9 = 6
6 + 10 = 16 - 10 = 6
6 + 11 = 17 - 11 = 6
6 + 12 = 18 - 12 = 6

7 + 0 = 7 - 0 = 7
7 + 1 = 8 - 1 = 7
7 + 2 = 9 - 2 = 7
7 + 3 = 10 - 3 = 7
7 + 4 = 11 - 4 = 7
7 + 5 = 12 - 5 = 7
7 + 6 = 13 - 6 = 7
7 + 7 = 14 - 7 = 7
7 + 8 = 15 - 8 = 7
7 + 9 = 16 - 9 = 7
7 + 10 = 17 - 10 = 7
7 + 11 = 18 - 11 = 7
7 + 12 = 19 - 12 = 7

8 + 0 = 8 - 0 = 8
8 + 1 = 9 - 1 = 8
8 + 2 = 10 - 2 = 8
8 + 3 = 11 - 3 = 8
8 + 4 = 12 - 4 = 8
8 + 5 = 13 - 5 = 8
8 + 6 = 14 - 6 = 8
8 + 7 = 15 - 7 = 8
8 + 8 = 16 - 8 = 8
8 + 9 = 17 - 9 = 8
8 + 10 = 18 - 10 = 8
8 + 11 = 19 - 11 = 8
8 + 12 = 20 - 12 = 8

9 + 0 = 9 - 0 = 9
9 + 1 = 10 - 1 = 9
9 + 2 = 11 - 2 = 9
9 + 3 = 12 - 3 = 9
9 + 4 = 13 - 4 = 9
9 + 5 = 14 - 5 = 9
9 + 6 = 15 - 6 = 9
9 + 7 = 16 - 7 = 9
9 + 8 = 17 - 8 = 9
9 + 9 = 18 - 9 = 9
9 + 10 = 19 - 10 = 9
9 + 11 = 20 - 11 = 9
9 + 12 = 21 - 12 = 9

10 + 0 = 10 - 0 = 10
10 + 1 = 11 - 1 = 10
10 + 2 = 12 - 2 = 10
10 + 3 = 13 - 3 = 10
10 + 4 = 14 - 4 = 10
10 + 5 = 15 - 5 = 10
10 + 6 = 16 - 6 = 10
10 + 7 = 17 - 7 = 10
10 + 8 = 18 - 8 = 10
10 + 9 = 19 - 9 = 10
10 + 10 = 20 - 10 = 10
10 + 11 = 21 - 11 = 10
10 + 12 = 22 - 12 = 10

11 + 0 = 11 - 0 = 11
11 + 1 = 12 - 1 = 11
11 + 2 = 13 - 2 = 11
11 + 3 = 14 - 3 = 11
11 + 4 = 15 - 4 = 11
11 + 5 = 16 - 5 = 11
11 + 6 = 17 - 6 = 11
11 + 7 = 18 - 7 = 11
11 + 8 = 19 - 8 = 11
11 + 9 = 20 - 9 = 11
11 + 10 = 21 - 10 = 11
11 + 11 = 22 - 11 = 11
11 + 12 = 23 - 12 = 11

12 + 0 = 12 - 0 = 12
12 + 1 = 13 - 1 = 12
12 + 2 = 14 - 2 = 12
12 + 3 = 15 - 3 = 12
12 + 4 = 16 - 4 = 12
12 + 5 = 17 - 5 = 12
12 + 6 = 18 - 6 = 12
12 + 7 = 19 - 7 = 12
12 + 8 = 20 - 8 = 12
12 + 9 = 21 - 9 = 12
12 + 10 = 22 - 10 = 12
12 + 11 = 23 - 11 = 12
12 + 12 = 24 - 12 = 12

Time:__________

Name:__________________________ Date:________________

Addition Practice

Say It! Write It! Repeat It!

1+1=2	1+2=3	1+3=4
1+1=2	1+2=3	1+3=4
1+1=2	1+2=3	1+3=4
1+4=5	1+5=6	1+6=7
1+4=5	1+5=6	1+6=7
1+4=5	1+5=6	1+6=7
1+7=8	1+8=9	1+9=10
1+7=8	1+8=9	1+9=10
1+7=8	1+8=9	1+9=10

Time:__________

Name:______________________ Date:______________

Addition Practice

Say It! Write It! Repeat It!

2+1=3	2+2=4	2+3=5
2+1=3	2+2=4	2+3=5
2+1=3	2+2=4	2+3=5
2+4=6	2+5=7	2+6=8
2+4=6	2+5=7	2+6=8
2+4=6	2+5=7	2+6=8
2+7=9	2+8=10	2+9=11
2+7=9	2+8=10	2+9=11
2+7=9	2+8=10	2+9=11

Time:__________

Name:________________ Date:________________

Addition Practice

Say It! Write It! Repeat It!

3+1=4	**3+2=5**	**3+3=6**
3+1=4	3+2=5	3+3=6
3+1=4	3+2=5	3+3=6
3+4=7	**3+5=8**	**3+6=9**
3+4=7	3+5=8	3+6=9
3+4=7	3+5=8	3+6=9
3+7=10	**3+8=11**	**3+9=12**
3+7=10	3+8=11	3+9=12
3+7=10	3+8=11	3+9=12

Time:________________

Name:____________________ Date:____________

Addition Practice

Say It! Write It! Repeat It!

4+1=5	**4+2=6**	**4+3=7**
4+1=5	4+2=6	4+3=7
4+1=5	4+2=6	4+3=7
4+4=8	**4+5=9**	**4+6=10**
4+4=8	4+5=9	4+6=10
4+4=8	4+5=9	4+6=10
4+7=11	**4+8=12**	**4+9=13**
4+7=11	4+8=12	4+9=13
4+7=11	4+8=12	4+9=13

Time:__________

Name:____________________ Date:____________

Addition Practice

Say It! Write It! Repeat It!

5+1=6	**5+2=7**	**5+3=8**
5+1=6	5+2=7	5+3=8
5+1=6	5+2=7	5+3=8
5+4=9	**5+5=10**	**5+6=11**
5+4=9	5+5=10	5+6=11
5+4=9	5+5=10	5+6=11
5+7=12	**5+8=13**	**5+9=14**
5+7=12	5+8=13	5+9=14
5+7=12	5+8=13	5+9=14

Time:__________

Name:__________ Date:__________

Addition Practice

Say It! Write It! Repeat It!

6+1=7	6+2=8	6+3=9
6+1=7	6+2=8	6+3=9
6+1=7	6+2=8	6+3=9
6+4=10	6+5=11	6+6=12
6+4=10	6+5=11	6+6=12
6+4=10	6+5=11	6+6=12
6+7=13	6+8=14	6+9=15
6+7=13	6+8=14	6+9=15
6+7=13	6+8=14	6+9=15

Time:__________

Name:____________________ Date:________________

Addition Practice

Say It! Write It! Repeat It!

7+1=8	**7+2=9**	**7+3=10**
7+1=8	7+2=9	7+3=10
7+1=8	7+2=9	7+3=10
7+4=11	**7+5=12**	**7+6=13**
7+4=11	7+5=12	7+6=13
7+4=11	7+5=12	7+6=13
7+7=14	**7+8=15**	**7+9=16**
7+7=14	7+8=15	7+9=16
7+7=14	7+8=15	7+9=16

Time:__________

Name:______________________ Date:______________

Addition Practice

Say It! Write It! Repeat It!

8+1=9	8+2=10	8+3=11
8+1=9	8+2=10	8+3=11
8+1=9	8+2=10	8+3=11
8+4=12	8+5=13	8+6=14
8+4=12	8+5=13	8+6=14
8+4=12	8+5=13	8+6=14
8+7=15	8+8=16	8+9=17
8+7=15	8+8=16	8+9=17
8+7=15	8+8=16	8+9=17

Time:______________

Name:______________________________ Date:________________

Addition Practice

Say It! Write It! Repeat It!

9+1=10	**9+2=11**	**9+3=12**
9+1=10	9+2=11	9+3=12
9+1=10	9+2=11	9+3=12
9+4=13	**9+5=14**	**9+6=15**
9+4=13	9+5=14	9+6=15
9+4=13	9+5=14	9+6=15
9+7=16	**9+8=17**	**9+9=18**
9+7=16	9+8=17	9+9=18
9+7=16	9+8=17	9+9=18

Time:____________

Name:______________________________ Date:____________________

Say It! Write It! Repeat It!

1+1=2 ____________________

____

____

____

1+2=3 ____________________

____

____

____

1+3=4 ____________________

____

____

____

1+4=5 ____________________

____

____

____

1+5=6 ____________________

____

____

____

1+6=7 ____________________

____

____

____

1+7=8 ____________________

____

____

____

1+8=9 ____________________

____

____

____

1+9=10 ____________________

____

____

____

1+10=11 ____________________

____

____

____

1+11=12 ____________________

____

____

____

1+12=13 ____________________

____

____

____

Time:__________

Name:______________________ Date:______________

Say It! Write It! Repeat It!

2+1=3	2+2=4	2+3=5
2+4=6	2+5=7	2+6=8
2+7=9	2+8=10	2+9=11
2+10=12	2+11=13	2+12=14

Time:__________

Name:________________________ Date:________________

Say It! Write It! Repeat It!

3+1=4

3+2=5

3+3=6

3+4=7

3+5=8

3+6=9

3+7=10

3+8=11

3+9=12

3+10=13

3+11=14

3+12=15

Time:__________

Name:______________________________ Date:____________________

Say It! Write It! Repeat It!

4+1=5	4+2=6	4+3=7
4+4=8	4+5=9	4+6=10
4+7=11	4+8=12	4+9=13
4+10=14	4+11=15	4+12=16

Time:________

Name:________________ Date:________________

Say It! Write It! Repeat It!

5+1=6 ____________

5+2=7 ____________

5+3=8 ____________

5+4=9 ____________

5+5=10 ____________

5+6=11 ____________

5+7=12 ____________

5+8=13 ____________

5+9=14 ____________

5+10=15 ____________

5+11=16 ____________

5+12=17 ____________

Time:________

Name:________________________ Date:________________

Say It! Write It! Repeat It!

6+1=7	6+2=8	6+3=9
6+4=10	6+5=11	6+6=12
6+7=13	6+8=14	6+9=15
6+10=16	6+11=17	6+12=18

Time:__________

Name:______________________ Date:______________

Say It! Write It! Repeat It!

7+1=8

7+2=9

7+3=10

7+4=11

7+5=12

7+6=13

7+7=14

7+8=15

7+9=16

7+10=17

7+11=18

7+12=19

Time:__________

Name:______________________ Date:______________________

Say It! Write It! Repeat It!

8+1=9

8+2=10

8+3=11

8+4=12

8+5=13

8+6=14

8+7=15

8+8=16

8+9=17

8+10=18

8+11=19

8+12=20

Time:__________

Name:________________________ Date:________________

Say It! Write It! Repeat It!

9+1=10 9+2=11 9+3=12

9+4=13 9+5=14 9+6=15

9+7=16 9+8=17 9+9=18

9+10=19 9+11=20 9+12=21

Time:__________

Name:________________ Date:________________

Say It! Write It! Repeat It!

10+1=11	10+2=12	10+3=13
10+4=14	10+5=15	10+6=16
10+7=17	10+8=18	10+9=19
10+10=20	10+11=21	10+12=22

Time:________

Name: ______________________ Date: ______________

Say It! Write It! Repeat It!

11+1=12 ______

11+2=13 ______

11+3=14 ______

11+4=15 ______

11+5=16 ______

11+6=17 ______

11+7=18 ______

11+8=19 ______

11+9=20 ______

11+10=21 ______

11+11=22 ______

11+12=23 ______

Time: __________

Name:______________________________ Date:__________________

Say It! Write It! Repeat It!

12+1=13	12+2=14	12+3=15
12+4=16	12+5=17	12+6=18
12+7=19	12+8=20	12+9=21
12+10=22	12+11=23	12+12=24

Time:__________

Darick Books

The First Bookstore of Its Kind

By Darick Spears

www.darickbooks.com

Made in United States
Orlando, FL
04 November 2026

90203579R00028